SPACE MYSTERIES
AND WONDERS 2020'S

INTRODUCTION

In my new book "Space Mysteries and Wonders 2020's", I showcase the latest big discoveries in space like a planet raining diamonds, a Super-Earth that could contain life, the mysterious Planet X and a star that's older than the Universe itself. We chronicle new, mysterious objects just spotted in space that scientists have never seen before, spectacular galaxies that shouldn't exist and the birth of baby stars and planets.

We also examine innovation in space technologies like the world's first Mars helicopter, the super-secret USAF Space Plane, DARPA's shapeshifting drone to explore distant worlds, solar sailing through space and NASA's developing technologies for a new space station and sustainable Moon bases.

My book takes you on an exciting journey of space discoveries, wonders, mysteries and explorations that you won't want to miss.

Ed Kane

TABLE OF CONTENTS

AUTHOR'S BIOGRAPHY

Ed Kane created and serves as Executive Producer of CEO Global Foresight, a national news program on PBS focused on breakthrough innovations. He is the author of 22 books on the latest innovations across industries. Ed is a science graduate of the University of Pennsylvania.

1. PLANET X – THE BIG PERTURBER IN SPACE

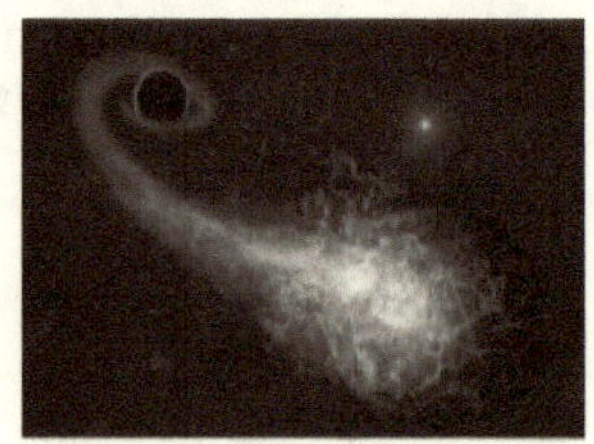
Source: NASA

Celestial Mystery on Edge of Solar System

The space phenomena on the outer edges of the solar system is known as Planet X, Planet 9 and Planet Next. And, global astronomers are baffled by it. They are not even sure that it's a planet. Some scientists speculate it's a grapefruit-sized extremely condensed and powerful Black hole. Its gravitational pull is so strong that it's also called the Big Perturber. Among the global astronomy community, opinion is divided. Some believe it is the new Planet X that is 5 to 10 times more massive than the Earth. Others think it is a powerful, extremely packed Black Hole with all of its explosive energy and mass inside a ball the size of a grapefruit. Quite a difference of scientific space speculation!

Tech to Determine by Late 2023

A powerful new piece of technology. - the highly anticipated Vera C. Rubin Observatory - is being built in the Chilean Andes. It is a massively powerful telescope that will go on line by the end of 2022. It has highly advanced technology that will be able to spot potential black holes, investigate dark energy and dark matter, find and track dangerous asteroids and study the Milky Way's formation and evolution. Several top astronomers at Harvard University are leading the exploration of Planet X. With the technology being deployed by the Rubin Observatory telescope, they expect to determine by late 2023 if the mysterious Planet X/9/Next/Big Perturber is actually a planet or an extremely powerful Black hole.

2. OBJECTS IN SPACE NEVER SEEN BEFORE

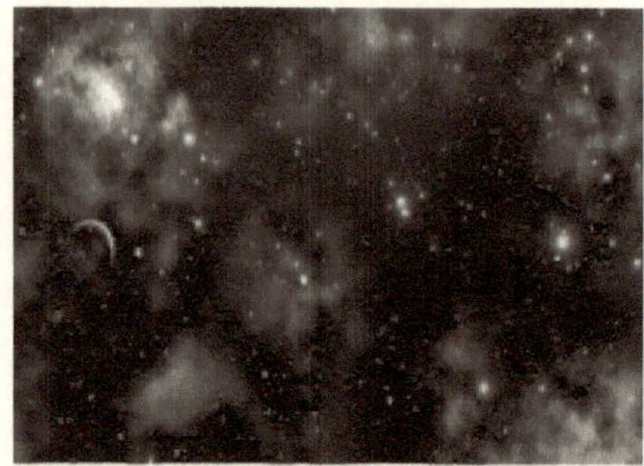

Source: Deep Space Image

Space: What's Up There?
Four mystery objects have been spotted by astronomers in deep space. The global team of astronomers say they've never seen anything like these objects before. Astronomers from the Royal College of Canada and Queens University, along with a team in Australia, discovered the objects using massive radio telescopes. They describe them as "edge-brightened discs", two are relatively close together and space objects "never probed before".

Deep Space Mystery
This discovery is one of the biggest mysteries in the universe. Astronomers are asking: What Are They? According to the team that discovered them: "We speculate they may be spherical shock waves from an extra-galactic event or a remnant from a radio galaxy viewed end-on". Whatever they are, they have the space world abuzz.

3. UK GOES INTO ORBIT

Source: OneWeb

OneWeb – Internet for Everyone, Everywhere
The United Kingdom and an Indian telecom giant are going into the satellite business together. The British government and Bharti Enterprises have bought OneWeb satellite company for $1 billion. The company was founded by US entrepreneur Greg Wyler with the purpose of providing internet everywhere to everyone with 648 low Earth satellites in orbit.

Rolling the Space Dice
This is a new lease on life for OneWeb and a gamble with potentially big payouts for the UK and Bharti. OneWeb was purchased out of bankruptcy by the UK and Bharti in July 2020. OneWeb filed for backruptcy late in March 2020 after its biggest financial backer SoftBank refused to provide any more funding. At the time of the filing, OneWeb had 74 satellites in orbit.

UK Mission and Goals
The UK is looking to add positioning technology to new satellites

in order to beef up the American GPS system it uses. It's also replacing bandwidth it is losing with the EU's Galileo network because of Brexit. There are estimates that the UK and Bharti will have to invest at least an additional $1 billion to bring OneWeb up to its full potential of providing continuous global internet service.

Crowded Space(s)

OneWeb is far from alone in the endeavor to provide universal internet through satellite constellations. SpaceX and entrepreneur Elon Musk are building a rival constellation network Starlink, now composed of about 600 satellites. When the number of OneWeb satellites hits 200, it may be able to go commercial. It's expected to first provide coverage at the North and South Poles for prospective military and gas and oil exploration customers. OneWeb is dual headquartered in the US and UK and needs to secure regulatory approvals for the purchase.

4. WORLD'S FIRST MARS HELICOPTER

Sourcee: NASA Ingenuity

Paired with Rover Perseverance

The world's first Mars helicopter - NASA's Ingenuity - will start operations in Mars' Jezero Crater in February 2021. It will be ferried there by NASA's Perseverance Rover. This is the first ever helicopter that will be flying in another world. It was developed by NASA along with Lockheed Martin and it is small, weighing just 4 pounds. The duo are expected to be launched from Cape Canaveral, FL aboard a rocket to Mars by late 2020. It's a great new mission of rocket science.

Delicate Maneuverings

Once established on Mars, the Perseverance-Ingenuity travel maneuvers will be delicate. Ingenuity will be attached horizontally under Perseverance for a hoped for February 18, 2021 landing at Jezero Crater on Mars. The landing clearance is tight. For 2 months, they'll stay together and hunt for a flat, unobstructed space to do test flying operations, at which time Perseverance will move football fields away.

Just the Right Spot

The decision-making is controlled by NASA experts on Earth. The right space for takeoff needs to be 33' x 33'. NASA operators will check all systems before Ingenuity flies. The process of unlocking the helicopter & making it completely vertical involves spectacular technology that even includes pyrotechnic fires releasing the legs. The helicopter will charge its batteries with solar panels.

30 Martian Days

Ingenuity is designed to work for 30 sols or Martian days to determine how feasible Martian helicopter flying will be for future missions. Ingenuity carries no instruments. It is providing the basis for the next generation of space helicopters for scouting difficult terrain, exploring deep craters and caves and even possibly carrying astronauts. As for Perseverance, the rover's main mission is searching for signs of ancient life on Mars and bringing samples back to Earth.

5. Center of the Solar System

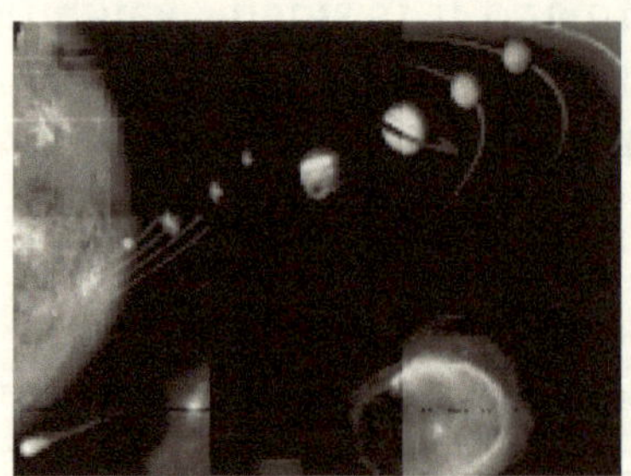

Source: Stock image of Solar System

International Astronomers Nail It

An international team of astronomers have narrowed down the center of the Earth's Solar System to within 328 feet. It is not dead center in the Sun as had been commonly thought. Instead, it's located just outside the surface of the Sun. The astronomers pinpointed the surprising location by monitoring the emissions of pulsars.

Pulsar Monitoring

This finding was not an easy task. There are a lot of gravitational forces influencing the calculation. The team used specially designed software to monitor pulsars. Pulsars are highly magnetized dead stars that emit electromagnetic radiation. Through the North American Observatory for Gravitations Waves, they were able to observe regular pulses from the pulsars and gravitational waves. Using that data, they were able to locate the center of the Solar System.

6. EU ACCELERATES SPACE RACE

Source: ESA Satellite

New Tech, New Satellites, New Innovations

The European Union is accelerating the investment of more money into its space explorations, satellite communications and rocket launches. Europe and the European Space Agency (ESA) have been very successful in space. They want to increase their efforts and keep pace particularly with the US and Chinese space programs. This, according to European Commissioner Thierry Breton in the summer of 2020.

EU Space Innovations

For the first time, the EU budget will be used to develop new technologies to launch rockets, including reusable rockets like SpaceX has developed. Breton has several novel ideas to spur space innovation:

- $1 billion euro Space Fund to support space startups
- $16 billion euros targeted just for space in the next EU

budget

- Competition to provide free access to launches and satellites for startups.

Bold Space Plan

This is a bold and innovative space plan for the EU. It includes rolling out a new generation of satellites in 2027 that interact with each other and provide more precise signals. It also includes the development of a new Space Traffic Management system to avoid collisions with the increasing number of satellites in space.

7. SUPER EARTHS DISCOVERED

Source: HARP Gliese 887 & Exoplanets

2 Exoplanets That Could Support Life

A team of German scientists has discovered two, Super-Earths just 11 light years away from Earth. The astronomers spotted them near the red dwarf star Gliese 887. They are outside of our solar system and they could sustain life.

Earth Like

University of Gottingen astronomer Sandra Saffers reported in the journal Science that the 2 exoplanets provide the best possibilities for the search for life outside of our solar system. In their report, the astronomers said they believe the exoplanets could contain liquid water and have a rocky surface like the Earth. They also have calculated that the surface temperature on these exoplanets at 158 degrees (F) or 70 degrees (C).

Gliese 887b & Gliese 887c

The newly discovered Super-Earths have been named Gliese 887b and Gliese 887c. NASA says more than 4,000 exoplanets have been discovered in space. Exoplanets have a mass larger than the Earth but smaller than the ice planets Neptune and Uranus. The German astronomers made their discovery using Chile's European Southern Observatory's High Accuracy Radial Velocity Planet Searcher (HARP) spectrograph.

8. VIRGIN GALACTIC, NASA AND ISS

Source: NASA

Virgin Galactic's Share Price Soars

Sir Richard Branson's Virgin Galactic and NASA are jointly developing a program for private missions to the International Space Station (ISS). Likely candidates for the private missions include scientific researchers, wealthy private citizens and space tourists. Virgin Galactic, like Elon Musk's SpaceX, is building a growing portfolio of business relationships with NASA.

Virgin Galactic's Role

Virgin Galactic will identify entities interested in buying private missions to the ISS. The cost is extremely pricey. A ticket to and from the ISS costs an estimated $58 million plus $35,000 per night onboard the ISS. The company will provide members of the private missions with astronaut training programs, transportation and necessary resources both while in orbit and on the ground as part of the preparations for the mission.

NASA's Strategy of Working with Private Companies

NASA is increasingly working with companies like SpaceX and Virgin Galactic as it prepares for its manned mission to the Moon in 2024. NASA's strategy is to reduce costs through the partnerships. In late May 2020 SpaceX successfully transported two US astronauts to the ISS. That was the first launch of US astronauts from US soil in more than a decade. NASA will debut a new space launch system for the journey to the Moon in 2021. NASA has big plans for a long-term presence on the Moon and a manned mission to Mars during the 2020's...all of which will be accomplished with strategic business partnerships.

9. STUNNING STAR FIREWORKS

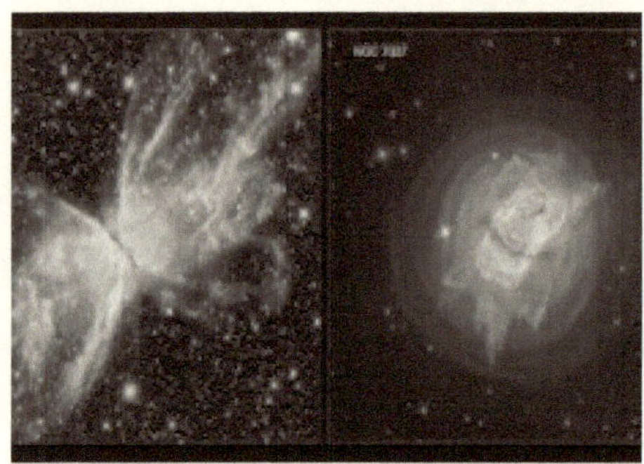

Source: Hubble Space Telescope

"Two Stars Going Haywire"

The NASA/ESA Hubble Space Telescope has captured new images of two nearby stars putting on a spectacular fireworks display in outer space. They are two planetary nebulae named the Butterfly nebula and NGC 7027. Astronomers say they are highly unusual, among the dustiest planetary nebulae ever discovered and contain unusually large amounts of gas.

Nuclear Fusion

Stars are engines of nuclear fusion. They live peaceful lives for billions of years. But near the end of their lives the fireworks start with erupting jets of hot gas from them exploding into the universe. Astronomers say the complexity and rapid changes in the jets of gas shooting off these two stars is unprecedented. The new Hubble images demonstrate that the two stars are splitting themselves apart in a very short timeframe.

Unprecedented Access

Hubble is providing astronomers with unprecedented access to this celestial event. They believe that NGV 7027 has been shooting off jets of its mass for centuries. But, according to lead researcher Joel Kastner: "Something recently went haywire in the center, producing a new cloverleaf pattern with bullets of material shooting out in specific directions." A team of astronomers led by Kastner is trying to pinpoint the cause of this spectacular, double star event.

10. SPACEX CREW DRAGON

Source: NASA - Crew Dragon Capsule

Many Historic Firsts on Earth and Space

In late May 2020, Elon Musk's SpaceX made history by carrying two US Astronauts to the International Space Station (ISS). This was the first time in nearly a decade that a rocket launched carrying US astronauts into space orbit from US soil. It was also a first for SpaceX to carry humans - NASA astronauts - into space. And it was the first time that a private commercial vehicle took US astronauts into orbit. Experts say this could be the beginning of a new era of human spaceflight for the US.

Historic Launch Time

The launch took place on May 31, 2020 from NASA's Kennedy Space Center in Cape Canaveral, Florida. President Donald Trump and Vice President Mike Pence were there. Two veteran US astronauts Doug Hurley and Bob Behnken crewed the SpaceX Crew Dragon capsule that rocketed into space on a SpaceX Falcon 9 Rocket.

Into Orbit

The Falcon 9 Rocket released Crew Dragon into a low Earth orbit 12 minutes after takeoff. During the next 19 hours, the astronauts orbited the Earth in Crew Dragon. They did some manual flying of Crew Dragon including raising the altitude of its orbit to move toward and reach the International Space Station.

Autonomous Docking with ISS

Crew Dragon has an autonomous docking system and didn't need much human help to connect and deliver the astronauts to ISS. Using a series of sensors and cameras to fly toward the ISS, it latched on to the open docking port of the ISS. Crew Dragon has successfully launched to the ISS before without any humans onboard. SpaceX has been developing this spacecraft and, along with NASA, testing it for six years.

Crew Dragon & Solar Panels

Crew Dragon operates with solar panels that last about four months in space. So the US astronauts will work in the ISS for a few months. When NASA decides it's time to return home to Earth, Astronauts Hurley and Behnken will get back into Crew Dragon, plunge back into the Earth's atmosphere and four huge parachutes will lower them into the Atlantic Ocean off of Florida. Crew Dragon has many emergency scenario, fail-safe systems to help protect the crew should there be a significant failure during the journey. This is rocket science, so we're wishing them the best of luck! The Falcon 9 Rocket is scheduled to return to Earth and land on a drone ship in the Atlantic Ocean when this historic journey is Mission Accomplished.

11. A GALAXY THAT SHOULDN'T EXIST

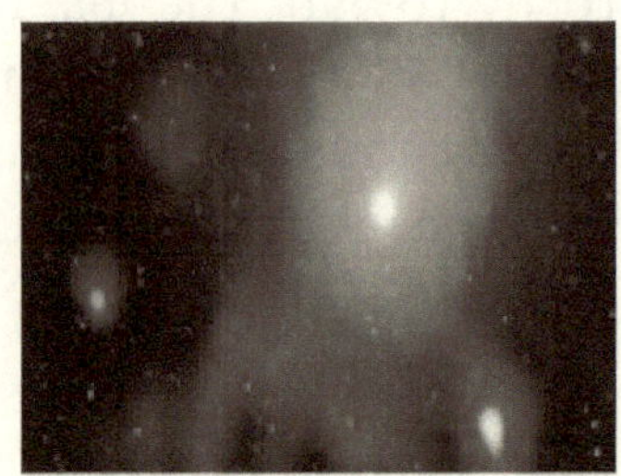

The Wolfe Disk

Astronomers have discovered a massive disk galaxy deep in the universe. It is almost as old as the universe itself at 12.5 billion years old. The universe is believed to be 13.8 billion years. It's called the Wolfe Disk and is yet another wonder in space just unveiled by astronomers, who are baffled by its very existence.

The Galaxy That Shouldn't Be

What's intriguing is that astronomers say the Wolfe Disk shouldn't exist and shouldn't be where it is in deep space. Those statements are based on astronomers' current understanding of galaxies. Their key question is: how did such a well formed rotating disk galaxy form at a time of chaos in space? A time when galaxies formed during violent collisions and chaotic mergers.

Created in a Different Way

The astronomers believe the Wolfe Disk was created and grew in a very different way. They theorize that it grew through the steady accretion of cold gas, which built its well-formed rotating disk. They're calling it the cold method of galaxy formation.

ALMA

Wolfe Disk is the most distant rotating disk galaxy ever observed. It was spotted by the Atacame group of massive telescopes, known as ALMA, in Chile. Wolfe Disk has a mass that is 70 billion times bigger than that of the Sun and rotates at 170 miles per second. It is quite a celestial find.

12. DISCOVERY OF CLOSEST BLACK HOLE TO EARTH

Source: European Space Agency

A Black Hole "In Our Neighborhood"

European astronomers have found the closest black hole to Earth ever discovered. It is so close that two stars circling it can be seen with the naked eye. The black hole is called the HR 6819 triple system. It was created by the death of a fleeting young star. Two remaining super-hot stars continue to orbit it.

Trillions of Miles Away From Earth

It is the closest black hole to Earth discovered thus far but it is still 1,000 light years away. Every light year equals 5.9 trillion miles. It was discovered by astronomers using the European Space Agency's massively powerful telescope at the European Southern Observatory in northern Chile. Lead astronomer Thomas Rivinius says the black hole "is in our neighborhood". He adds there may be 100 million to one billion of these relatively small but hugely dense objects in the Milky Way.

Black Holes Where Nothing Escapes

The difficulty of discovering black holes is nothing escapes from them. Even light doesn't escape so they are virtually invisible. In the case of this new discovery, astronomers found the black hole because of the unusual orbit of the two stars.

Dating Back 15 Million Years

This black hole was created 15 million years ago, according to the astronomers, when one of the three stars became too hot, too big, transformed to a supernova and then turned into a black hole in a violent process. The astronomers say the black hole is four to five times the mass of the Sun and its accompanying stars are three times hotter than the Sun.

13. NASA, VIRGIN GALACTIC AND SUPERSONIC TRAVEL

Source: NASA-Virgin Galatic

Flying Public at Supersonic Speeds

NASA is teaming up with Sir Richard Branson's Virgin Galactic space company to develop high speed technologies for supersonic air travel. The purpose is to make high Mach speed vehicles for the travelling public here on Earth to get from "Point-to-Point" cities on Earth in vastly reduced times. NASA and Branson finalized their collaboration by signing a Space Act Agreement.

Vision and Goal

Branson's vision for transporting the public at Mach speeds is a "global network of spaceports, transcontinental supersonic space flights and delivering passengers anywhere in the world in a couple of hours". To achieve this, the aircraft will actually be spacecraft.

Technological First

The envisioned SST would be a spacecraft that would launch into space from its departure point, then re-enter the Earth's atmosphere and arrive at the destination point for landing. The flight would happen in a fraction of the time a conventional plane would take. This spacecraft would be a technological first. The takeoff, space flight, re-entry, landing and reuse as a high speed civilian transportation system have never been achieved.

SSTs That Respect the Environment

Branson's deal with NASA brings his supersonic travel vision many steps closer to reality. And Branson says the supersonic spacecraft that he and NASA aim to develop will be environmentally responsible and sustainable. No timetable was offered as to when the new collaboration will start delivering new supersonic vehicles. But this concept brings space travel right down to Earth for global city to city travel.

14. HUMANITY'S FIRST LOOK AT BIRTH OF A PLANET

Source: European Space Agency

Creation of a Planet Deep in Space

Astronomers have discovered a new planet being born inside a swirling orange disc surrounding a distant young star. And, the team of international astronomers who discovered it says this is the first time that humanity has witnessed the birth of a planet. They add this discovery will help them to better understand how planets come to exist around stars.

Celestial Creation

As you can see in the picture, the new planet is forming in the very bright yellow twist in the inner region of the orange, swirling disc that surrounds the young star AV Aurigae. It was spotted by the European Space Agency's European Southern Observatory's Very Large Telescope in northern Chile.

520 Light Years Away

The birth of the new planet is taking place 520 light years from Earth. Lead astronomer Anthony Boccaletti of the Observatoire de Paris in France says the image shows cosmic matter at a gravitational tipping point collapsing and forming a new world. The international team of astronomers are from France, the US, Taiwan and Belgium. The image is the deepest observation of the AB Aurigae system to date.

15. SUN'S GRAND SOLAR MINIMUM

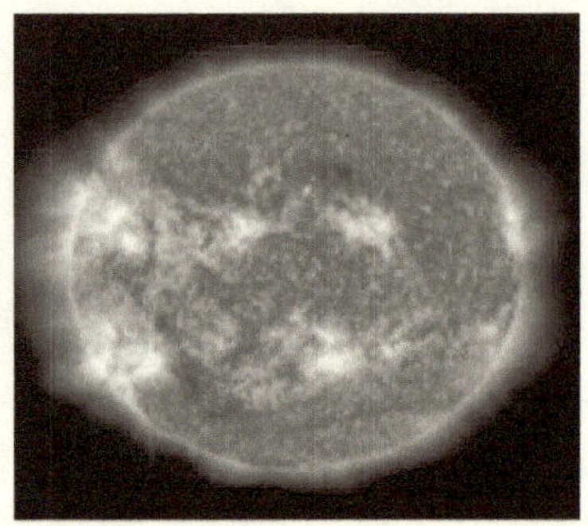

Source: NASA

Start of a New Ice Age?

NASA scientists say that the Sun is in a much less active phase, meaning far less solar flares, energy and sunspots. The phase is called a Grand Solar Minimum, which hasn't occurred since 1650 to 1715, when it triggered a Little Ice Age in the Northern Hemisphere with lower temperatures. The question is: Are we headed toward another Ice Age?

Climate Change

NASA scientists say we are not on the brink of a new Ice Age. And the new scientific basis of their conclusion is interesting. They say the warming of the Earth from fossil fuels and their greenhouse gas emissions is six times greater than even decades long cooling from the Grand Solar Minimum. In fact, NASA adds: "Even if the Grand Solar Minimum were to last a century, global temperatures would continue to warm." A shocking measure of the impact of global warming.

Solar Dynamics

The Sun is very dynamic and goes through 11 year cycles of peak and low solar activity, solar flares, energy and sunspots. The center of our Universe is ever changing.

16. 2020'S COMET SENSATIONS

Cosmic Snowball

Comet SWAN is a cosmic sensation. Forbes magazine calls it The Comet of the Year. And, it was a visible phenomena to many in 2020. Comets aren't rare in our skies, but this one is. The reason: it was so exceptionally bright and beautiful to view. It put on some significant light shows especially in the Southern Hemisphere.

Comet Neowise

Comets are cosmic snowballs of frozen gas, rock and dust that orbit the Sun. Another comet sensation Neowise thrilled star-gazers in mid-summer 2020. Its brilliant display of light won't be visible to Earth again for another 6,800 years. NASA says the last icy space wonderer to put on such a dazzling show was Hale-Bopp back in 1997. Neowise was so bright that it provided astron-omers with a lot more and better data than most comets do. The scientists believe that the data they've gathered from multiple telescopes will provide key details about the composition and structure of the 3 mile wide comet Neowise.

17. MEGA SPACE TRAVEL TREND - RIDESHARING

Source: Loft Orbit

Small Sats, Low Cost, Low Earth Orbit

Ridesharing is a new megatrend in Space. A SpaceX Falcon 9 rocket will fly on a ridesharing mission, late in 2021, carrying a Loft Orbiting satellite with a DARPA Blackjack Pit Boss mission system inside for a space demonstration. Loft buys space buses (satellites) and rents space. The US Defense Department's Advanced Research Projects Agency (DARPA) will be flying onboard the SpaceX rocket with many other customer payloads. DARPA's Blackjack programs aims to deploy low cost, small, low earth orbit satellites into operation for the US Military.

Space Experiment

DARPA's testing of Blackjack will commence in early 2021. The experiment is called Sagittarius A. It's designed to demonstrate any risks or problems before the main Blackjack satellites go into production and launch in late 2021.

Pit Boss

Pit Boss is the software mission management system that will enable the satellites to autonomously acquire, process and provide information to the system users. Sagittarius A demonstration will test the flight computer and an optical imaging sensor that searches for targets in the ocean. The data goes into the satellite's flight computer, AI processes it and then redirects the satellite to start searching for more targets. This real, dry run in space is expected to usher in the deployment of Blackjack in 2022. It will also contribute to increased "Ridesharing" to space.

18. SUPER-SECRET USAF SPACE PLANE

Source: MIT Technology Review

Launched May 2020 and On Duty

The US Space Force & USAF's top secret mission of its unmanned, robotic, long duration, X-37B space plane is underway. It hurtled into space in May 2020 onboard a United Launch Alliance Atlas V rocket for what is now being called the US Space Force's OTV-6/USSF-7 mission. This is another secret mission for one of the most fascinating space planes in the universe.

Green, Solar Powered, Space Plane

The X-37B is an unpiloted, long distance, long duration flight spacecraft. It flies classified missions that can last months and even years for the US military. It is a green machine that's powered by a solar array that's the size of a pickup truck and embedded inside the space plane.

Tremendous Technology

The technology is incredible. There are 2, X-37B space planes built by Boeing. The winged aircraft can land themselves back on

Earth, are reusable and look like a small NASA space shuttle. Plus, they are solar array powered. They have been on six missions since 2010 with 2,865 days in orbit.

Mission Experiments Onboard

The current mission for X-37B, like all the other missions, is classified and top secret. But it is known to contain several interesting experiments, that include a small Falcon Sat-8 satellite invented by the USAF Academy, NASA space environment experiments and a novel solar power beaming experiment from the Naval Research Laboratory.

19. NASA'S GROUND RULES FOR LIVING ON THE MOON

Source: Stock Moon image

NASA's Artemis Accords

NASA has released a set of basic principles designed to govern how humans will live and work on the Moon. They are called the Artemis Accords and include the main tenets of what is designed to become an international agreement for the exploration of the Moon. It has sparked international debate. Russia is angry that it's not part of the early negotiations. In fact, the Russian space agency Roscosmos has accused the Trump Administration of "planet snatching" in its efforts to drive formation of the Accords. China too has complained that it isn't part of the early negotiations.

Lunar Safety Zones and Mining Rights

The Accords are designed to establish safety zones around lunar bases to prevent what NASA calls "harmful interference". They would permit companies and nations to own the lunar resources that they mine. The Accords are part of NASA's strategy to win support from allies for its plans to build long term bases on the Moon as part of NASA's Artemis lunar mission.

International Space Law

At this point, the Trump Administration is bypassing the United Nations and attempting to craft bi-lateral and multilateral agreements on the Moon accords with allies. President Trump wants the Accords to establish under international law the rights of companies to mine the Moon and own those resources. Some Russian space experts have argued those lunar resources are the property of all humanity.

20. JOURNEY TO THE SUN: NASA AND ESA'S BLACKBIRD MISSION

Source: NASA Launch of Solar Orbiter

Unprecedented Solar Orbiter Probe

NASA and the European Space Agency have successfully launched the Solar Orbiter on a journey to the Sun, to take unprecedented views of its blazing solar poles. The Solar Orbiter took off on top of an Atlas 5 rocket. The Orbiter separated from the rocket within 53 minutes to go into space flight as planned and is now in communications with NASA. All of this happened in February 2020. This unique mission is going very well as it moves along on a very long journey to the Sun.

Mission Unprecedented

This US and European mission is unprecedented. It's expected to help scientists understand how the Sun's massive amount of energy affects humans in space and how it impacts all of us on the Earth. The mission will provide unique images, views and data of the sun's blazing pole regions, which are critical sources of data.

Ten Year Journey

The orbiter is equipped with awesome equipment including an array of solar panels and antennas. The journey to the Sum will take ten years. The orbiter will position at 26 million miles from the Sun, which is 95% of the distance between the Earth and the Sun. It will map the Sun's poles which have a concentrated source of solar winds. Those winds are high impact. They penetrate our atmosphere and even can impact satellites.

Main Goal

NASA and ESA scientists want to determine how the Sun creates and controls the heliosphere, which is the massive bubble of protection that surrounds the solar system. They want to know why the bubble changes over time. NASA and ESA scientists believe the answer may be found in the Sun's poles. Those views will first be available from the probe in 2025. Exciting new science for all of us to observe from Earth!

Key Goals to Get This Off the Ground

The key to this mission, that was first proposed in 1999, was to develop a thermal protection system that can withstand the intense heat of the sun. The Orbiter was built in Europe. It's a great example of US-European space exploration. The cost of the Solar Orbiter hurtling into space is $1.5 billion.

21. VIRGIN GALACTIC'S AIR TRAVEL REVOLUTION

Source: Virgin Galactic

LA to Tokyo in 2 Hours

Billionaire entrepreneur Sir Richard Branson is a brilliant innovator. And he has beyond supersonic plans for his space travel company Virgin Galactic. Sometime in 2021, he expects to have flown the first commercial tourism flight into space. Nearly 8,000 tourists have signed up for a space flight at a cost of $250,000 each. But Branson business plans go way beyond space tourism. He wants to revolutionize commercial air travel by using his space vehicles for greatly accelerated air travel here on Earth.

Beyond Supersonic Speed - Spaceline Flying

Virgin Galactic says it is the only company that can deliver flying at greater than supersonic speed in a winged aircraft. It calls the high speed flying "Spaceline" flying. The company says its spaceship Unity and two other space vehicles under construction can be used to vastly accelerate long distance fly times, for instance from New York to Paris. The company pinpoints the market potential for this type of flying at $900 billion. Just capturing the premier business flyer portion of the market would be worth billions.

High Speed Global Mobility

Virgin Galactic's business plans for growth are strategic and aggressive. Besides space tourism, it plans on offering extremely high speed air travel that could cut the fly time between LA and Tokyo to two hours. As part of that, Virgin Galactic will develop what it calls "high speed mobility vehicles". Also it's pursuing commercial and government users of its proprietary technologies.

22. SNOWMAN STAR BORN BY TWO STARS MERGING

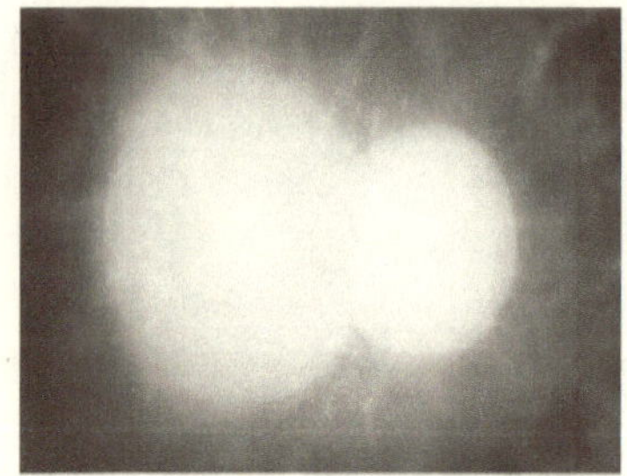

Source: Warwick University, UK

Unique Shape and Unique Atmosphere Composition

Nothing like this new star has ever been discovered in space before. Warwick University scientists have spotted a huge, snowman-shaped star. It's a new star born by the merger of two "white dwarf" stars. They've named the star WDJ0551+4135. The star's atmosphere is unique and unusually rich in carbon. Everything about this star has the potential of leading to new scientific discoveries about the universe and the birth and death of stars.

Bigger Than the Sun

The new Snowman star is more massive than the Sun. Another unusual aspect of the birth of this star is that the merger of the two "white Dwarf" stars didn't result in an explosion and powerful supernova. The Warwick University scientists say what is particularly exciting for their research is that the stars merged rather than exploded, as is usual.

How a Star is Born

The scientists believe that measuring the properties of the failed supernova will tell them a lot about the "pathways to thermonuclear self annihilation" of stars. A "white dwarf" is what stars like the Sun become after using up all of their nuclear fuel. In this case, a new star was born by the successful merger of two starts.

23. NASA SPACE TOURIST DESTINATION FOR YOU

Source: Axiom

NASA Partners with Axiom

Space exploration company Axiom of Houston, Texas is part-
nering with NASA for a big launch of space tourism. Tourists
will be able to visit the International Space Station (ISS) as early
as 2024. Axiom is a private US space station manufacturer that
is building, what it hopes will be, the world's first commercial,
international space station. CEO Michael Suffredini is the former
program manager of the ISS for NASA. This is a big business alliance, with tremendous expertise behind it, for launching tourists
into space.

Advanced Luxury Space "Hotel" Priced as High as $35,000

The Axiom space vehicle is an impressive piece of technology.
First of all, it will be led by a trained astronaut, have a crew and
also research labs on board. It has the biggest space observation

window ever built into a space capsule and the state-of-the art interior was designed by the top French-based hotel and yacht designer Philippe Starck. This takes space tourism to a whole new level.

Space Nest

The interior is designed like a "nest" according to Starck with hundreds of nanoLEDs that change color and embedded touchscreens and handles. It will also have Wifi. Starck said the space provided him as a designer and the future occupants what he calls "multidimensional freedom" as it is zero-gravity with no horizontal, vertical or diagonal limitations.

Ticket Takes Money & Fitness

The Axium-NASA module is designed for fully trained astronauts whose nations aren't members of the ISS and for private citizens. If you can pay the price, you also need to be very fit to get a seat. Travelers to space must clear a physical and then 15 weeks of very rigorous training that includes jet flights, extreme environment endurance and suborbital space flights. If you clear, you get one ticket to ride.

24. BEYOND THE BIG BANG

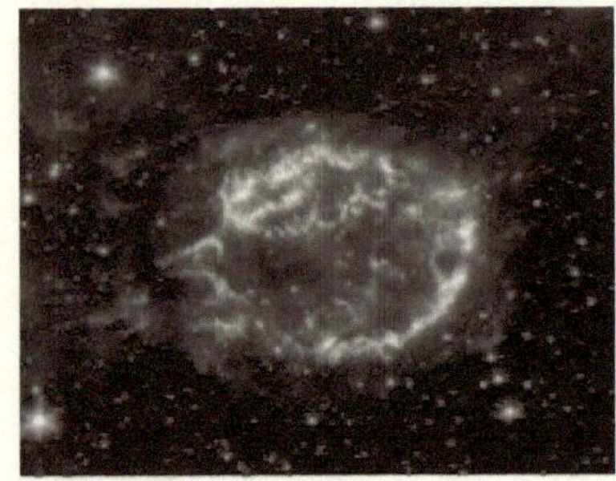

Source: NASA

Huge Explosion in Universe From Light Years Ago Detected
Astronomers using data from the most advanced NASA, European Space Agency, Indian and Australian telescopes have discovered the biggest explosion that the universe has ever experienced since the Big Bang. It happened in a galaxy 390 million light years away from the Earth. And incredibly it took place in slow motion over hundreds of millions of years. The science to discover this phenomenom is remarkable and the details emerging on events going back into space from such a distance and time are amazing, innovative scientific discoveries.

Supermassive Black Hole in the Ophiuchus Galaxy Cluster
The global team of astronomers say the explosion they detected released five times more energy than ever detected before. The explosion came from a supermassive black hole in a galaxy called the Ophiuchus Galaxy Cluster nearly 400 million light years from

the Earth. The power of the explosion was enough to punch a hole into the gas surrounding the black hole that is big enough to contain 15 contiguous Milky Way galaxies. This is an amazing scientific discovery through breakthrough telescopic innovation. What it means to the formation of the Universe and Earth as we know it is the next chapter of the history of the Universe.

25. NASA DISCOVERS MARS IS A DYNAMIC PLANET

Source: NASA

Hundreds of Marsquakes in the Past Year

NASA's robotic lander InSight delivered some big discovery and innovation dividends. InSight detailed, for the first time, that the Red Planet Mars experiences quakes like the Earth and the Moon. NASA has called the shake, rattle and rolls on Mars "Marsquakes". Incredibly, InSight has detected 450 seismic events on Mars in the past year as it has explored Mars on the ground. 20 of the quakes were significant and in the 3 to 4 point magnitude range.

NASA First

According to NASA, this is the first time that they've established that Mars is a seismically active planet. InSight's observations will help scientists better understand how rocky planets like Mars and the Earth form and evolve. And the lander's sensors also detected significant winds swirling around Mars. In fact, there are thousands of passing whirlwinds whipping around Mars.

Mars Quakes

Compared to quakes on the Earth and the Moon, the Mars quakes are relatively small. Mars is more seismically active than the Moon but less so than the Earth. The Marsquakes provide NASA scientists with new data and information on the interior core of Mars. NASA says the quakes on Mars are caused by the long-term cooling of the planet that makes it contract and facture.

Tech Wonder

InSight is a robotic technological wonder. The lander is equipped with heat flow probes to take the planet's temperature, sensors to gauge wind and air pressure, seismometers to detect quakes and a magnetometer to detect magnetism.

26. STARDUST DISCOVERY OLDEST ON EARTH

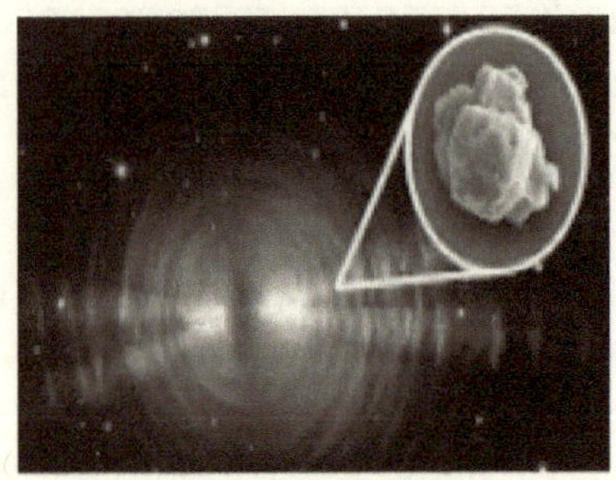

Source: NASA Meteorite

Stardust Material That's Older than the Sun

This is an extraordinary scientific discovery announced by a team of astronomers led by University of Chicago Professor Dr. Philipp Heck. In a meteorite that fell from space to Australia 50 years ago, there is awesome stardust dating back 5 to 7 billion years. This stardust predates the Sun. Heck and his team analyzed the contents of the meteorite. They found it to be the oldest, solid material ever found on Earth. Their work has been published in the prestigious Proceedings of the National Academy of Sciences.

What This Discovery Opens Innovative Views On

The scientists say this stardust will tell us how parent stars were formed in our galaxy. It will tell us about the origins of the oxygen that we breathe and provide the opportunity to trace materials back before the creation of the Sun. The stardust is the oldest material discovered to reach the Earth. The material is older than our solar system. The Earth is estimated to be 4.5 billion years old. The Sun is 4.6 billion years old. The newly found and analyzed stardust is 5 to 7 billion years old. This discovery does give a whole new perspective on our brilliant and beautiful universe!

27. SATURN'S TITAN MOON AND LIFE

Source: NASA Jet Propulsion Lab

Moonscapes, Science and Life

NASA scientists have mapped Saturn's exotic, largest Moon Titan. Apparently, Titan has quite a story to tell about the possibilities of life. NASA has unveiled the first global geological map of Titan. It shows a strange and exotic world that NASA scientists say is a strong candidate for the search for life beyond Earth.

Incredibly Interesting Terrain

The map shows dunes of frozen organic material, vast stretches of plains, and lakes and seas filled with liquid methane...all of which may harbor forms of life beyond Earth. The map is based on radar, infrared and data from NASA's Cassini spacecraft that studied Saturn and its moons from 2004 to 2017. The data and images now emerging are fascinating.

What's Next

Clearly, Titan contains organic material that is the ingredient critical to form life. What's next is NASA is going to launch its Dragonfly Mission sometime in the 2020's to reach Titan by 2034 with a multi-rotor drone to explore this exciting new Moon in space.

28. SPACEX UNVEILS MARS VEHICLE

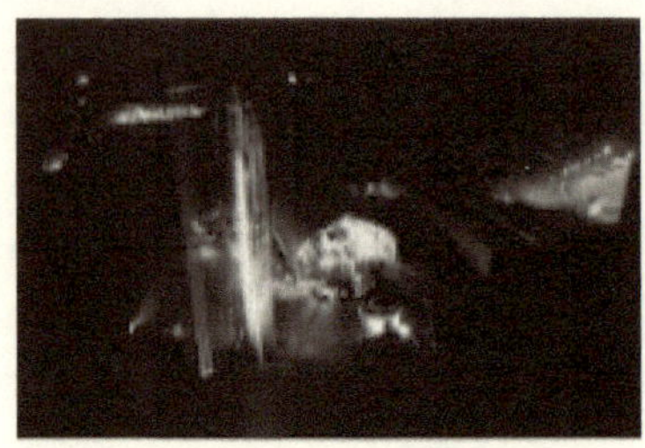

Source: SpaceX Starship

CEO Elon Musk on to Mars

SpaceX CEO Elon Musk unveiled his silver bullet: a spacecraft to take you to Mars and beyond. It's the new Starship Mk1 prototype, unveiled at SpaceX's South Texas test site in Boca Chico, TX. Starship is a massive, reusable launch system. When finalized, the vehicle will have a 387 foot Starship Super Heavy stack and be powered by six Raptor engines. The spacecraft has room for 37 Raptor engines, depending on the mission needs and distance.

Musk, a Visionary Thinker

Starship will be capable of taking up to 100 people back and forth to space, including the Moon, Mars and other space destinations. Musk says space travel has to be done like air travel. He believes that is the first breakthrough needed to make humans a space-faring, multi-planet civilization. He adds making space travel like air travel is "the fastest path to a self-sustaining city on Mars", which is his ultimate goal and destination.

Passengers in the 2020's

Musk says the Starship vehicle system could soar into space and back to Earth three times a day. And, if all goes well, people could start flying as passengers on the vehicle sometime during the 2020's

29. NASA'S SHAPESHIFTER EXPLORING NEW DISTANT WORLDS

Source: NASA/CalTech Concept

Source: NASA

Drone of Drones

NASA's Jet Propulsion Lab at Caltech has introduced the drone of drones. It separates into two different units: two shapes to explore new distant worlds and new habitats. It's a prototype and is made of several small, quadcopter drones called cobots. The cobots have propellers and fly independently and the system reassembles to roll on the ground. The future for the concept is much bigger. The Shapeshifter would be made up of a number of small robots that can easily self-assemble into larger robots and disassemble as the mission requires, particularly in space. The mini-robots will be able to fly, roll, float and swim and then morph into a single machine.

Morphing Robots

On the ground, the cobots come together to form a roller-wheel like drone to explore the ground in places in space like Titan, where there is very limited information about the surface. The unpredictability of the surface makes versatility and shapeshifting essential in the drone. In the future, the plan is to make the cobots work together as a team of twelve to explore caves, underwater areas and various types of terrain, including in outer space. The ultimate morphing robotic team would be carried aboard a mothership lander that would house their energy source and scientific instrumentation for testing and analysis.

Dragonfly 2026

This extraordinary technology will take quite a few years to fully develop. But a deployment target date for it is 2026 when NASA's Dragonfly drone takes off for Titan with possibly the Shapeshifter onboard. The Shapeshifter is a transformational vehicle to explore, treacherous, distant worlds.

30. HOW TO GENERATE LIGHT AT NIGHT

Source: UCLA

Renewable and Complements Solar Power
This new, inexpensive thermoelectric device may be transformative in energy generation. It harvests the coldness of space during the night to generate electricity - enough right now to power a LED light at nighttime, but the inventors say it's very scalable. The device is a significant new innovation from engineers at UCLA and Stanford University. The gadget works at night when solar systems don't. The inventors say it's a new approach to power generation when power at night is needed. It complements solar power that doesn't work at night, giving a 24/7 approach to green, renewable energy.

Phenomenon Like Frost Formation
The device takes advantage of radiative cooling, the process by

which frost forms on grass during above freezing temperatures at night. The sky facing surface of the technology passes heat to the atmosphere as thermal radiation. It loses some heat to space and reaches a temperature cooler than the surrounding air. That temperature differential produces renewable energy at night, when lighting demands are peak.

Scalable Tech for Global Use
According to the UCLA and Stanford engineers, their invention is highly scalable. The radiative cooling device essentially consists of an aluminum disk coated with paint and all the other components are readily available for purchase off the shelf. This is important innovation to watch for because of its practicality and scalability for worldwide use to supplement solar energy at night.

31. PLANET RAINING DIAMONDS – NASA DISCOVERY

Source: MIT

Discovery Unveiled

NASA scientists have discovered that Saturn's mysterious atmosphere is raining diamonds down onto the planet. This discovery is extraordinary and yet another example of the vast mineral deposits in space that commercial space ventures are targeting.

Saturn Rings

While studying how and when Saturn's rings of ice and dust emerged, the NASA scientists stumbled upon a surprising feature of Saturn's atmosphere. The atmosphere is composed of high amounts of sulfur, along with hydrogen and helium. The atmosphere is extremely harsh and unforgiving. So much so it rains diamonds onto the planet. NASA's Cassini probe made the shocking discovery when taking a closer look at Saturn's weather in 2013. The discovery was unveiled in a recent BBC documentary.

From Soot to Diamonds

Saturn is the sixth planet from the Sun and the second largest in the Solar System. According to scientist Dr. Brian Cox, Saturn's atmosphere is brutal. There are huge clouds filled with water and lightning that is 10,000 times stronger than on Earth. The combination changes methane gas in the atmosphere into huge clouds of soot. The pressure is so great the chunks of soot transform into diamonds, which rain down on the planet as liquid because of the intensity of Saturn's environment.

32. SUPERNOVA FOUND IN ANTARCTIC SNOW

Source: Stock Image Antarctic

Interstellar News

Australian scientists have discovered large amounts of stardust in Antarctic snow, likely from the explosion of a supernova that had been close to the Sun. The explosion rained down particles of a unique iron isotope discovered in melted snow from the South Pole. The discovery of the rare isotope, iron-60, which is not native to Earth, was made by a scientific research team from Australian National University.

South Pole Supernova

The researchers have ruled out any chance that the isotope was created by human activity. They say the only explanation is that it was created by an interstellar rock hitting the Earth. The team melted more than 1000 pounds of snow and analyzed the contents of the melt, which contained large amounts of isotope iron-60. Their next focus is to determine the age of the isotope and to dig deeper for more samples in the icepack. Their remarkable findings were published in the journal Physical Review Letters.

33. UNPRECEDENTED COSMIC EVENT

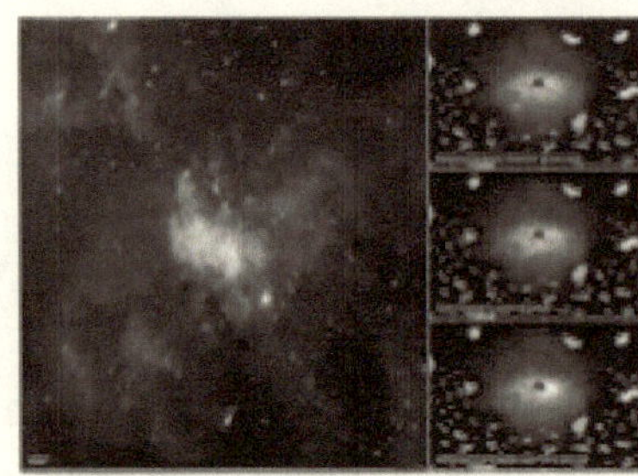

Source: NASA Sagittarius A

Sagittarius A: Black Hole Light Show
The black hole Sagittarius A (Sgr A) is the Earth's closest black hole and is right in the middle of the Milky Way. Using the Keck II Telescope in Hawaii, astronomers watched the massive black hole erupt and spew a huge burst of infrared radiation. They call the event "unprecedented" and they can't explain exactly what caused it.

Off the Charts
The astronomers say the black hole reached much brighter "flux levels" than ever observed before. According to NASA, SgrA is 26,000 light years from Earth. The astronomers disclosed it in the journal Astrophysical Journal Letters. According to the astronomers, black holes are always variable but this one was off the charts of historic data including from the Keck II Telescope.

Theories
There are two theories as to what caused the huge burst of light. A

star passing by could have changed the gas flow around the black hole. Or it could have been a flash from a passing gas cloud.

34. MARTIAN ROCK TELLS BIG STORY

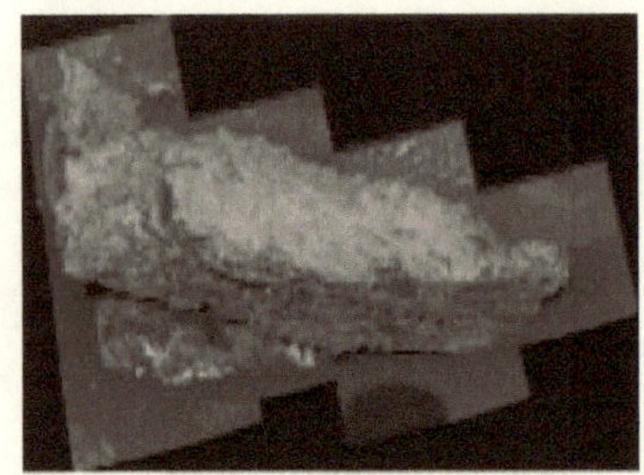

Source: NASA

Shows Presence of Water, Raises Questions of Life on Mars
NASA's Curiosity rover has been exploring the planet Mars for the past seven years. It found a rock that contains some of the secrets to what Mars was like when it was a lot more wet. The discovery is so important that the rock has been nicknamed "Strathdon".

Strathdon
The rock was found in the Gale Crater or the "clay bearing unit". The rock is composed of layers of sediment that are unique and wavy. NASA scientists say this shows evidence of flowing water and blowing wind. They add the history of water on Mars is much more complicated than they had previously thought. They now do not believe that Mars went from wet to dry overnight.

Signs of Life on Mars
The question still remains: with the presence of water on a warmer Mars with a thicker atmosphere, could that have sustained more than microbial life? NASA is searching for that an-

swer.

35. STEAM-POWERED TWIN SPACECRAFTS

High Theater in Low Earth Orbit

Two cubesats (small satellites) powered by steam have successfully performed a complicated and coordinated maneuver in space. With the help of NASA on the ground, one of the spacecraft commanded the other to close the 5.5 mile gap between them. It did so successfully.

Old Fashioned Steam Power

The twin spacecrafts' fuel tanks are filled with water. Thrusters convert the water into steam which propels the cubesats. NASA says the maneuvers demonstrate how small spacecraft can work together on future missions. NASA adds these spacecraft are designed to operate in swarms. These space teams could be deployed in the future for deep space exploration and work autonomously to explore new worlds. NASA plans on sending cubesats to the Moon in the 2020's.

36. STAR THAT'S OLDER THAN THE UNIVERSE

Source: NASA's Hubble Shot of the Star

Star Discovered That May Be Older Than Our Universe

It's a very old star discovered by astronomers that's a cosmic riddle of intergalactic proportions. Astronomers have found a star that they think may be older than the universe. Best estimates by astronomers is that the universe is 13.8 billion years old. But they have found a star close to earth that is estimated to be 14.5 billion years old, based on its very low metal content. The star is HD 140283, nicknamed the Methuselah Star. It's raising a lot of questions, fascination and wonder. Is that beautiful twinkling star older than time and the universe itself?

The Methuselah Star

The questions come from many including a top physicist at the UK's Royal Astronomical Society Dr. Robert Mathews. He calls it a riddle of cosmic proportions. How can the universe contain stars older than itself? This question sparks questions about how accurate calculations are about the age of the universe. Even NASA's estimate is imprecise: Methuselah could be 800 million years younger or older than 14.5 billion years. It is too early to tell but Dr. Mathews is looking into new research into gravitational waves as a possible explanation for the age differential. But, he adds the age of the universe question is back with a vengeance.

37. CHARTING THE MILKY WAY

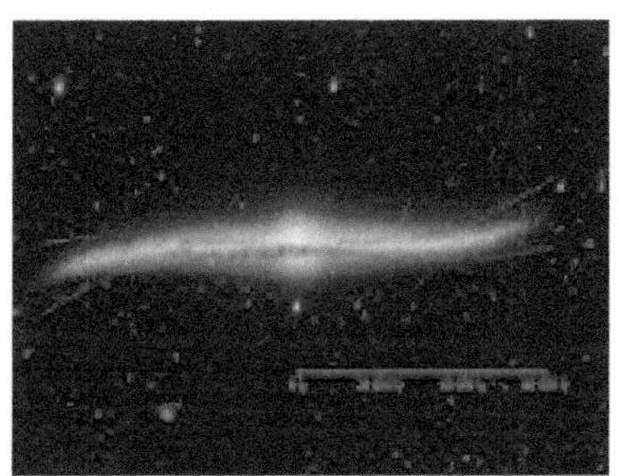

Surprise Finding "Warped and Twisted", Not Flat

Polish astronomers from the University of Warsaw have created the most precise and largest map of the Milky Way to date. By tracking thousands of blinking stars in the galaxy, they've created a 3D scale map of the system. They discovered that it isn't flat as a pancake but twists and turns in shape much more than previously thought by astronomers.

Spiral Galaxy

The Milky Way is a spiral galaxy measuring 120,000 light years across. It's a mass of stars, gas and dust about 27,000 light years from the Earth. The Polish team found the distortions in the Milky Way are immense with some stars 60,000 light years away from the Milky Way's center. And, the galaxy's thickness is variable.

Potential Causes

The astronomers cited several potential causes for the variations including interaction with nearby galaxies, intergalactic gas and even dark matter. Their groundbreaking work was published in the journal Science.

38. SUPER EARTH FOUND IN SPACE

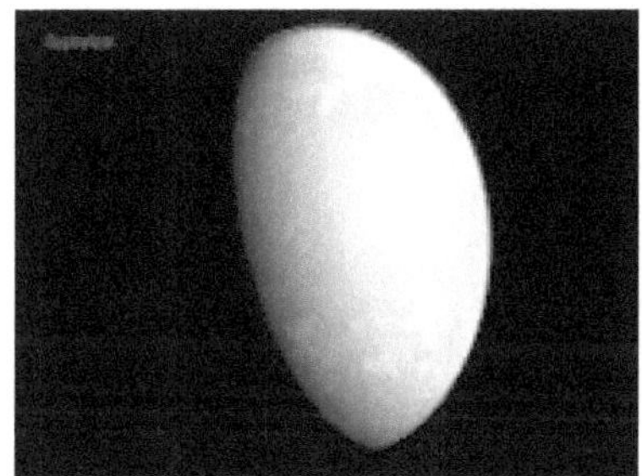

Source: NASA's Goddard Space Flight Center

May Have "Earth Like" Conditions

Astronomers have discovered what they're calling a Super Earth 31 light years away from our solar system. They say it is "potentially habitable". The planet is named GJ 357d. It is six times larger than the Earth and it orbits a sun that is much smaller than ours. An international team of scientists discovered it using powerful ground telescopes and following up on data first relayed by NASA's planet hunting satellite TESS. The astronomers believe that it could provide "Earth like conditions" for life.

Looking for Signs of Life

Highly advanced telescopes will, in the next several years, be going on line that will be able to pick out any signs of life on Super-Earth. Two telescopes going live in 2021 and 2025 should show whether the planet is rocky and if it has any oceans. If the planet's atmosphere is thick, it could support water on the surface and sustain life. If there is no atmosphere, the average temperature would be 64 degrees below zero, making it more glacial than habitable.

39. SOLAR SAILING THROUGH SPACE

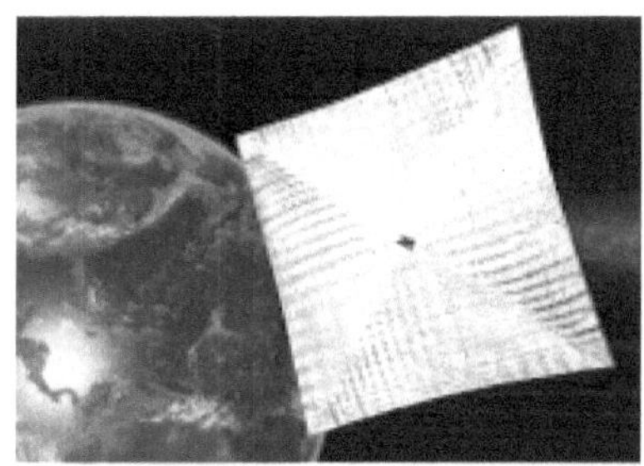

Source: The Planetary Society's LightSail in Orbit

Solar Powered Space Missions

This green energy technology has the potential to be disruptive innovation. It's a solar powered sail invented by the non-profit Planetary Society in California, led by the Science Guy Bill Nye. The LightSail 2 CubeSat spacecraft successfully deployed its solar sail in space. It's the first spacecraft to be propelled by sunlight alone. This is very important science because the use of solar sails to convert solar wind into thrust could save fuel and provide clean solar power for spacecrafts on long space missions.

Solar Sailing

The small spacecraft is the size of a toaster. It was hurled into space aboard a Space X Falcon Heavy rocket in the summer of 2019. Flight controllers on the ground in California fully deployed the solar sail from the satellite into space. The sail is the size of a boxing ring and the system is orbiting the Earth solely on the power of the sun. This system is designed to test solar sailing technology and works very well. It's in full orbit and has relayed pictures back to Earth.

Photon Power

The LightSail 2 is a square composed of four triangles and made of aluminized Mylar. When the sunlight's protons hit the Sail, the pressure creates solar wind and pushes the craft forward. Light-Sail has completed its original year long mission in space and has entered an "extended space mission" phase to optimize the solar sailing control software and the vehicle's space performance.

40. NASA'S 21 WHOLE NEW WORLDS

Source: NASA

Planet Hunter Extraordinary

NASA's planet hunter TESS spacecraft - Transiting Exoplanet Survey Satellite – recently marked its first year in space and the results are extraordinary. The spacecraft has discovered 21 new worlds outside of our solar system. It's also spotted 850 possible exoplanets that have to be confirmed and 6 supernovas. For NASA the results are way beyond all expectations.

Star Search

TESS is focusing on stars that are less than 300 light years away. It looks for dips in brightness which indicates that an object just passed across the star. The data will help NASA determine which exoplanets it wants to explore in-depth and where life might exist.

Historic Mission

This is the most comprehensive search for planets ever accomplished by mankind. TESS finished exploring the southern part of the sky and is examining the northern sky.

41. SUSTAINABLE LUNAR BASES

Source: Stock Image of Future Lunar Bases

Fascinating Research by the European Space Agency
Administrators of the European Space Agency (ESA) believe that the next step in space exploration is building lunar bases for astronauts. To do that a sustainable source of energy is required to sustain human life and power for rovers and landers. ESA scientists think that they have one, using the surface of the Moon. ESA scientists have created bricks composed of lunar regolith, which is the soil, dust and rocks on the surface of the Moon collected on past lunar missions. In testing, the bricks are able to store solar energy and generate it for use as power, heat and electricity. It's possible this innovation could provide future lunar bases a sustainable energy source.

Scaling Up the Technology
The ESA team is now scaling up the process and efficiency of their heat energy bricks from lunar regolith. They say that travelers to the Moon, with this new energy source, wouldn't have to take

much with them from Earth. They also think it has the potential to enable very ambitious missions into space. As the world marked the 50th anniversary of the Apollo 11 historic mission to the Moon, the work being done by the ESA showcases humanity's continuing quest and ingenuity to journey into space and form habitats there.

42. SPACEBOT FROM SWITZERLAND

Source: ETH Zurich

Dynamic Walking & Flight Phases

The Spacebot, a quadruped robot, was created by students at the ETH Zurich and ZHAW Zurich, Switzerland. It's designed to travel in low gravity environments like the Moon, asteroids and Mars. It travels by "Dynamic Walking" similar to the travel methods used by the original lunar explorers, the Apollo 11 crew, who used a hopping gait in the low gravity environment of the Moon, 50 years ago.

Dynamic Walker

The bot travels by leaping. All four legs leave the ground and leap to heights over six feet. This allows for full flight phases with all legs off the ground. The lower the gravity, the longer the flying time. The Spacebot is being tested at the European Space Agency (ESA). The Swiss team says the robot basically behaves like a mini-spacecraft. They add, this type of dynamic walking is more efficient and saves energy in low gravity environments like Mars and the Moon.

Dynamic Bot
The robot has spring-loaded legs that store energy and releases it for the next leap. It also has a reaction wheel similar to that used in satellites. The wheel enables the robot to take off and land safely and get acclimated on the ground for exploring. The next step is to take Spacebot out of the ESA test labs in the Netherlands where lunar gravity is simulated and into the real world to take on challenging terrain.

43. ROBOT TO INSTALL TELESCOPES ON THE MOON

Source: Stock image of Moon

Telescopes to Look Deep Into the Universe

A NASA funded lab at the University of Colorado is developing robots to deploy small, highly advanced telescopes on the far side of the Moon. In the next ten years, the NASA team will send a rover aboard a lunar lander spacecraft and place it on the dark, far side of the Moon.

Humans & Machines Working Together

The plan is to set-up a network of small telescopes. The deployment will be performed by the rover's robotic arm. The arm will be controlled by astronauts in an orbiting lunar station called Gateway. Gateway will serve as transit to and from the Moon and as a refueling station for deep space missions.

New Telescopic Views of Deep Space

On the far side of the Moon, the telescope will be free of light and noise. That will enable a unique and pristine gaze into the deep reaches of space. It's a leading edge example of projects underway by NASA, private companies like Space X and other nations. These missions will open up and change the landscape of the Moon forever.

44. DISTANT DISCOVERY IN SPACE

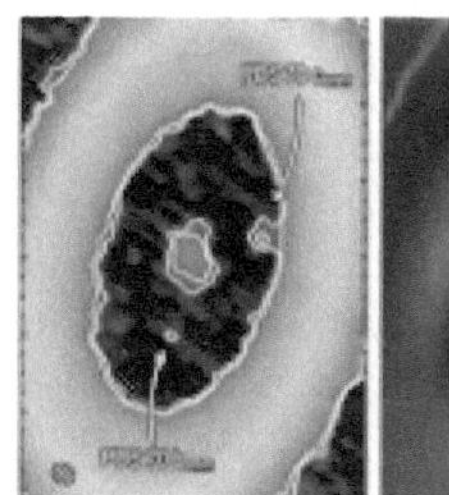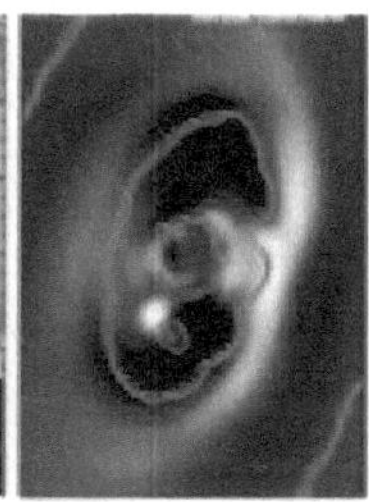

Source: Rice University

World First

Astronomers from Rice University have spotted a moon-forming disk around a very distant exoplanet. The disk, as seen in the color enhanced millimeter wave radio signal on the left, is made of gas and dirt. Scientists say it is similar to the one that is believed to have created the moons around Jupiter four billion years ago. This is a first time observation of a moon forming disk.

World's Most Powerful Telescopes

The Rice University team made the discovery by using the most powerful array of telescopes on Earth. They are located at the ALMA Observatory in Chile.

Exoplanet PDS 70c

The exoplanet is named PDS 70 c. It is a still forming gas giant that's located 370 light years from the Earth.

45. NASA PLANS NEW SPACE STATION

Source: NASA Lunar Orbiter Concept

Man on the Moon in 2024

NASA's plan is for a mini-space station that would orbit the moon with astronauts onboard for 30 to 60 day stays as they travel back and forth to the surface of the Moon. The new space station will be a lunar orbiter, lander and astronaut habitat. It's called the Lunar Orbital Platform - Gateway that NASA wants to use to speed travel and exploration of the Moon. The Trump Administration wants humans on the Moon again by 2024.

Gateway on a Fast Track

NASA wants the 1st phase of Gateway - the Power & Propulsion phase - to launch in 2022. The 2nd phase is the lunar habitat where the astronauts will live, work and do scientific experiments, on and near the Moon. In 2024, it's expected that both components will be launched on a commercial rocket.

International Partners

Canada has already signed on as a partner on the new space station with NASA. Canada is going to build a robotic arm called Canadarm 3. The European Space Agency is also looking to partner. The space station will be equipped to enable the astronauts to do spacewalks. And, NASA envisions that at a later date it will have commercial uses and also be a stopover for astronauts on their way to Mars.

46. SPACE CYBERSECURITY AT RISK

Source: European Space Agency

Royal Institute of International Affairs Report

UK based Chatham House or The Royal Institute of International Affairs has released a report warning that there is an urgent need to address the cybersecurity of satellites in space. They are asking NATO and NATO member nations to address the cybersecurity of space-based satellite control systems because they are vulnerable to cyberattack, particularly if a network system were breached.

International Security

There are important ramifications from this cyber vulnerability finding because virtually all military operations rely on space-based data and communications for instant decision making. According to the research report, any threat to a satellite's control system or available bandwidth "poses a direct challenge to national critical assets" and international security.

Critical Space Assets

With the military dependent on space based architecture, Chatham House researchers say there are new and growing cyber threats that can put military missions in jeopardy. The group is calling for a major investment to harden and protect satellites from hacking so that they can provide reliable and accurate information particularly for the military.

47. NASA SPOTS ELECTRIC SOCCER BALLS IN SPACE

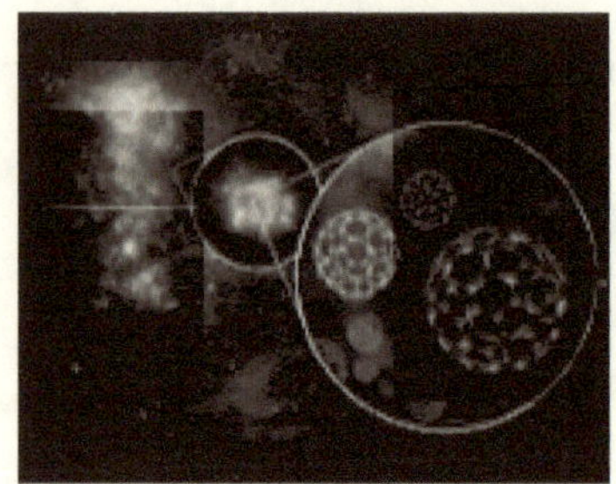

Source: NASA/JPL - CALTECH

Interstellar Presence

The Hubble Space Telescope has discovered electrically charged molecules shaped like tiny soccer balls in space. They say this is highly significant because it sheds light on the mysterious contents of gas and dust that fill interstellar space, the so-called interstellar medium (ISM). This is the first time that an electrically charged, ionized "buckyball" has been found in the interstellar system.

Buckyballs

The buckyball molecules are composed of 60 carbon atoms that configure like a soccer ball. The molecules are a form of carbon called buckminsterfullerene or Buckyballs after the inventor of the geodesic dome, the inventor Buckminster Fuller. NASA scientists say this discovery sheds light on ISM and the beginning of planets.

ISM

NASA scientists believe that interstellar gas and dust are the starting point of the chemical process that creates stars and planets. Life as we know it is based on carbon bearing material. Now that NASA scientists have discovered that carbon bucky-balls have formed and survived in the harsh environment of interstellar space, they want to determine how widespread this might be in the universe.

48. OUT OF THIS WORLD EXOPLANET

Source: NASA Artist Rendering

Exoplanet LHS 3844b

A NASA space telescope has revealed for the first time a rocky Earth-sized exoplanet. It's outside of our solar system and tightly orbits around the most common type of star in our home galaxy, the Milky Way. The exoplanet has no atmosphere to sustain life.

Atmospheric Void

Researchers published their discovery in the journal Nature. They say the surface of the exoplanet, known as LHS 3844b, is most likely barren like the Moon or covered in dark volcanic rock. The exoplanet is 49 light years away from the Earth. It's one of 4000 exoplanets, discovered in the past 20 years, that orbit stars in the Milky Way. It's 1.3 times larger than the planet Earth.

49. UAE'S HOPE MARS MISSION

United Arab Emirates Going to Mars

The United Arab Emirates has big space ambitions and plans. It is going to launch a mission to Mars in the late summer of 2020, making it the first Arab nation to launch an interplanetary mission. The Hope Mars probe will deploy the Hope Satellite to study the climate on Mars.

Destination Mars 2021

When the Hope satellite reaches Mars in 2021, it will show how Mars' climate conditions change throughout the year. It will provide a "holistic" mapping of Mars and orbit the Red Plane for two years. The mission could be extended through 2025. Interesting, a female scientist is a key player behind the mission. Sarah Al Amiri is deputy project manager of the Mars Mission and Chairperson of the UAE Council of Scientists. 90% of the workers on the Emirates Mars Mission are under 35 years old.

Highly Advanced Technology

The Mars probe will carry three highly advanced pieces of equipment. There's the Emirates Exploration Imager, which is a camera that will send high-resolution images back to Earth. The Emirates Mars Infrared Spectrometer will study ice, water vapor, dust and temperature patterns in the Martian atmosphere. And, the Emirates Mars Ultraviolet Spectrometer will study Mars' lower and upper atmosphere and be used to determine what causes hydrogen and oxygen on Mars to escape into space.

Crowded Martian Skies
The UAE Mars team is working with space experts at the University of California Berkeley, the University of Colorado Boulder and Arizona State University. Mars in the 2020's will be a crowded place with four major space missions underway. NASA's 2020 Rover mission, the European Space Agency's ExoMars rover mission, China's Mars Explorer and now the Emirates Mars Mission will all be probing Mars, which is 140 million miles from the Earth.

"SPACE MYSTERIES & WONDERS 2020's" © BY EDWARD KANE ON AMAZON

Copyright 2020 © Edward Kane

ALL RIGHTS RESERVED

recording or by any information storage or retrieval system or otherwise without prior written permission of the publisher. No warranty liability whatsoever is assumed with respect to the use of the information contained herein. Although every precaution has been taken in the preparation of this book, the publisher and author assume no responsibility for errors or omissions. Neither is any liability assumed for damages resulting from the use of the information contained herein. amazon.com/author/ekane

BOOKS BY JOURNALIST EDWARD KANE

Non-fiction books on innovations and discoveries across industries in 2020 and 2019.

TOP INVENTIONS FOR THE 2020's
Just Published by Amazon and Kindle amazon.com/author/ekane

SMART DEVICES FOR THE 2020's
Just published by Amazon and Kindle amazon.com/author/ekane

FUTURE TRAVEL VEHICLES

Just published by Amazon and Kindle amazon.com/author/ekane

FUTURE TRAVEL VEHICLES BY EDWARD KANE

SPACE

- "Space 2020's: What's Up There?" - Kindle & paperback
- "Bargain Space Trips" - Kindle & paperback
- "Space Renaissance in the 21st Century" - Kindle & paperback
- "Search For Life in Space" - Kindle & paperback

TRANSPORTATION AND TRAVEL INNOVATIONS

- "Future of Transportation: 2020's and Beyond" - Kindle & paperback
- "Electric Vehicles for All" - Kindle & paperback
- "Hot Electric Vehicles for the 2020's" - Kindle & paperback
- "Important Innovations: Transportation" - Kindle, paperback & Audiobook on Audible
- "How to Travel in the Future" Vol. 1 & 2 - Kindle & paperback

LISTS OF TOP NEW INNOVATIONS

- "List of Best New Innovations" - Kindle & paperback
- "List of Top New Environmental Innovations" - Kindle & paperback
- "List of Top New Gadgets" - Kindle & paperback
- "List of Top New Medical Innovations" - Kindle & paper-

back

- "List of Top New Energy Innovations" - Kindle, paperback & Audiobook on Audible
- "List of Top New Robots" - Kindle, paperback & Audiobook on Audible
- "How to Use AI & AR" - Kindle & paperback

INVESTING IN INNOVATIONS

- "Investing in Disruptive Innovations" - Kindle & paperback

Fiction - Adventure, Life Lessons Book for Children

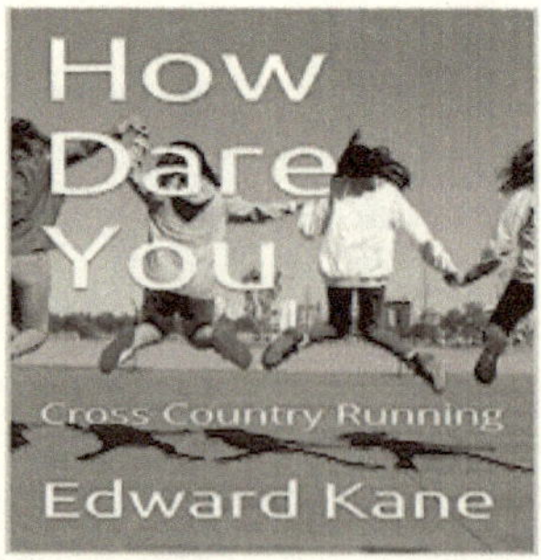

- "How Dare You" - Kindle, paperback & Audiobook on Audible

amazon.com/author/ekane

9 798666 732457